COPY 2

522
DAR

Darling, David J.

The new astronomy

$10.95

DATE			

© THE BAKER & TAYLOR CO.

THE NEW ASTRONOMY

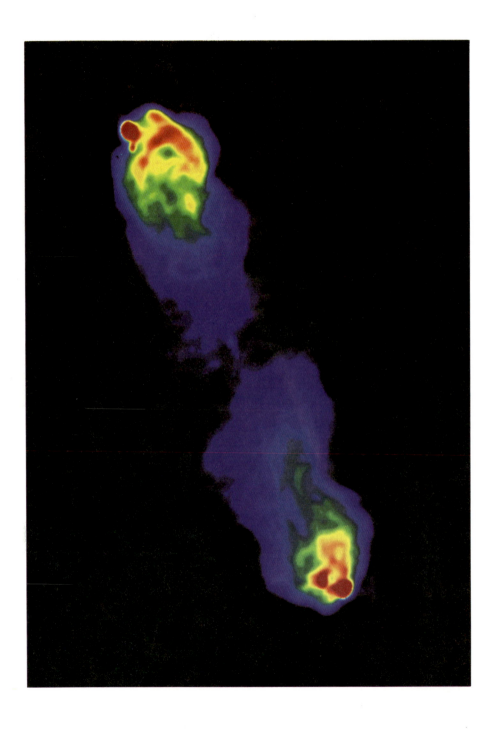

THE NEW ASTRONOMY
AN EVER-CHANGING UNIVERSE

by David J. Darling

Illustrated by Jeanette Swofford

DILLON PRESS, INC.   MINNEAPOLIS, MINNESOTA

Photographs are reproduced through the courtesy of AT&T—Bell Telephone Laboratories; Brookhaven National Laboratory; California Institute of Technology; Cerro Tololo Inter-American Observatory; Hale Observatories; Jet Propulsion Laboratory, California Institute of Technology; Kitt Peak National Observatory; the National Aeronautics and Space Administration; and the National Radio Astronomy Observatory, operated by Associated Universities, Inc., under contract with the National Science Foundation (observers: Alan H. Bridle, Richard A. Perley, Peter A.G. Scheuer, Robert A. Laing).

Dillon Press, Inc., 242 Portland Avenue South
Minneapolis, Minnesota 55415

Printed in the United States of America

Library of Congress Cataloging in Publication Data

Darling, David J.
     The new astronomy : an ever-changing universe.

     Bibliography: p.
     Includes index.
     Summary: Examines the state of modern
astronomy including X-ray, Gamma ray, infrared,
and UV astronomy, and discusses future possibilities
in the field.
     1. Astronomy—Juvenile literature [1. Astronomy]
I. Title.
QB46.D28   1985   522   84-23083
ISBN 0-87518-288-7

    4   5   6   7   8   9   10   91   90   89   88   87

Contents

Modern Astronomy Facts

World's Largest Telescopes

Telescope Name	Size of Mirror (inches)
Bol'shoi Teleskop Azimutal'nyi	236
George Ellery Hale Telescope	200
William Herschel Telescope	180
Multiple Mirror Telescope	177
Inter-American Observatory 4-meter Telescope	158
Anglo-Australian Telescope	153
Nicholas U. Mayall Reflector	150
United Kingdom Infrared Telescope*	150
Canada-France-Hawaii Telescope	142
E.S.O. 3.6-meter Telescope	141
Space Telescope	138

*Also used for optical astronomy

Location	Date of Completion
Mount Pastukhov, U.S.S.R.	1976
Palomar Mountain, California	1948
La Palma, Spain	1985
Mount Hopkins, Arizona	1979
Cerro Tololo, Chile	1976
Siding Spring, Australia	1975
Kitt Peak, Arizona	1973
Mauna Kea, Hawaii	1979
Mauna Kea, Hawaii	1979
Cerro La Silla, Chile	1976
Earth orbit	1986

Modern Astronomy Facts

World's Largest Radio Telescopes

Single Instruments

Location	Size of Dish (feet)
Arecibo, Puerto Rico	1,000
Efflesburg, West Germany	328
Jodrell Bank, England	250
Parkes, Australia	210

Interferometers

Location	Size of Dish (feet)	Number of Dishes
Socorro, New Mexico (Very Large Array)	82	27
Hooghalen, The Netherlands (Westerbork Synthesis Telescope	82	14
Cambridge, England (Five Kilometer Telescope)	43	8

Questions & Answers About Modern Astronomy

Q. If we think of the electromagnetic spectrum as a musical scale, how many "octaves" does it cover?

A. From the longest radio waves to the shortest gamma rays, about 60 octaves! (Note: each additional octave represents a halving of the wavelength.)

Q. Are there any objects in space that give off all kinds of electromagnetic waves at the same time?

A. Yes. Several quasars, active galaxies, and supernova remnants shine in all parts of the electromagnetic spectrum.

Q. What is the largest optical telescope in the world?

A. The reflecting telescope on Mount Pastukhov in the Caucasus Mountains (USSR). Completed in 1976, it has a mirror 19½ feet (6 meters) in diameter, weighing more than 40 tons. In the future, there are plans to build much larger instruments with mirrors more than 32 feet (10 meters) in diameter.

Q. What will be the main advantages of the Space Telescope?

A. First, it will be able to see objects in space about 50 to 100 times fainter than any seen before. Second, it will be able to make out much more detail than any telescope on the ground.

Q. Why must the important parts of an infrared telescope be made very cool (with a substance like liquid helium)?

A. Everything at room temperature, including our bodies and telescope equipment, gives off invisible heat waves, or infrared. In order for these waves not to overpower the tiny amount of infrared reaching us from objects in space, the detector of an infrared telescope must be made as cold as possible.

Q. If we had "infrared eyes," what would we see if we looked up at the night sky?

A. Mostly we would see warm, interstellar clouds of gas and dust, together with the coolest of bright stars, such as red giants and supergiants. Many stars that seem bright in ordinary light would look quite dim to infrared eyes.

Q. What are "bursters"?
A. Points in the sky that send out sudden, short bursts of high-energy radiation. Both X-ray and gamma ray bursters have been found. They are thought to be caused by matter falling onto the surface of a dense star such as a white dwarf or a neutron star.

Q. Do ordinary stars give off X rays?
A. Yes. Most kinds of stars give off some X rays from superhot gas in their coronas, or outermost layers. X rays from the sun were first discovered in 1948 during a short rocket flight.

Q. What is "cosmic background radiation"?
A. Steady radiation that comes, with equal strength, from all parts of the sky. Astronomers have found background radiation at microwave, X-ray, and, possibly, gamma ray wavelengths.

WINTER AT KITT PEAK NATIONAL OBSERVATORY

1 Signals from Space

High in the sky on a clear, fall evening, you can see the bright star, Vega. Bluish-white in color, its surface sizzles at a temperature of more than 18,000°F (10,000°C). Measuring 2,700,000 miles (4,300,000 kilometers) in **diameter,*** Vega is about three times bigger than our own sun. Its distance from earth is 152 **trillion** miles (245 trillion kilometers).

Through a big telescope, you can see the remains of a star that blew up nine centuries ago. This is the **Crab nebula,** an object more than 200 times farther away than Vega.

Today, the hot gas of which the Crab nebula is made is still shooting out into the surrounding space. At its center is a **pulsar**—an old, burnt-out star, so squashed that a bucketful of its **matter** would weigh 100 trillion times more than a bucketful of water. Spinning dizzily 30 times a second, the Crab pulsar measures only about 20 miles (32 kilometers) in diameter but is heavier than the sun!

These are just a few of the amazing things we have learned about the universe around us. Yet our robot probes have explored only as far as the nearby planets. Astronauts have gone no farther than the moon. How can

*Words in **bold type** are explained in the glossary at the end of this book.

White light enters this picture at the upper left, passes through a prism, and splits up into all the colors of the spectrum.

we know so much about objects that lie trillions and trillions of miles away?

The answer is that everything in space—a star, a **galaxy,** even a cold cloud of gas—sends out signals. These signals are of various types and carry vast amounts of information. Traveling far through space, some eventually reach the earth. The best known space signals are those in the form of light that we can actually see with our eyes.

Secrets of a Light Beam

In 1666 Isaac Newton, the great English scientist, made a surprising discovery. On passing sunlight through a glass triangle—a **prism**—he found that light is split up into the colors of the rainbow. What seems like plain white light is really the colors from red—through

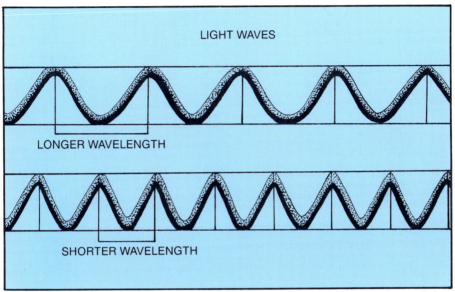

LIGHT WAVES

LONGER WAVELENGTH

SHORTER WAVELENGTH

Light is a wave motion made of electromagnetic waves. This drawing shows light waves with a longer wavelength (above) and a shorter wavelength (below).

orange, yellow, green, blue, and indigo—to violet, all mixed up together.

Later, scientists found that light is a wave motion, similar to ripples on the ocean. In the case of light, though, the waves are rather special. They are **electromagnetic waves,** made of ripples of electricity and magnetism.

Like all waves, light waves can be measured by their **wavelength.** This is the distance between one wave crest and the next. Changing the wavelength of an electromagnetic wave also changes its energy. The shorter the wavelength, the greater its energy becomes.

Red light, for example, has a longer wavelength and a lower energy than green light. The whole spread of colors, or **spectrum,** of light, covers wavelengths from 31 millionths of an inch (red light) to 15 **millionths** of an inch (violet light).

By measuring the strength of light at different wavelengths from a distant star, we can learn the star's temperature and **density.** We can discover what it is made of, too. From other measurements, we can find how fast the star is moving away from us or towards us. We can learn about any **magnetic field** it may have.

Signals other than light are also given off by objects in space. Beyond the red end of the visible spectrum, and beyond the violet end, there are other types of electromagnetic waves. To scientists, each has its own story to tell.

What Our Eyes Cannot See

At wavelengths longer than red light, there are **in-**

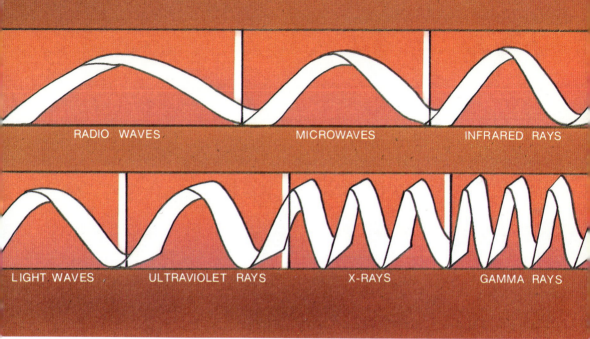

The various waves that make up the electromagnetic spectrum are shown here. Starting at the top left, radio waves, microwaves, and infrared rays have wavelengths longer than red light. On the bottom, ultraviolet rays, X rays, and gamma rays have wavelengths shorter than violet light. In between these two groups of invisible wavelengths is the visible light with which we are most familiar.

frared rays. These we can sense with our skin, rather than with our eyes. They are heat waves, which are given off by anything warm.

At still longer wavelengths, **radio waves** travel through space. Stretching over a huge wavelength range, from many miles to less than an inch, radio waves are used on earth for radio and television broadcasting, and for radar. They are the weakest form of electromagnetic radiation.

Ultraviolet rays and **X rays** travel at wavelengths shorter than violet light. Finally, at the shortest wavelengths and greatest energies of all, there are **gamma rays.**

All of these different types of electromagnetic radiation reach us from space. Each offers an important way of learning more about the universe. And yet, until recently, we were only able to make use of the visible light signals.

The Shield of the Atmosphere

Two big problems had to be solved before we could start to explore the universe at "invisible" wavelengths. First, special instruments were needed. The **radio telescope,** used to pick up radio waves from space, is an example.

Second, the earth's atmosphere stops many of the invisible waves from reaching the surface. The atmosphere allows light and radio waves to pass through. On the other hand, it blocks out completely all X rays and gamma rays, along with most ultraviolet and infrared. To see them, scientists must send instruments above the atmosphere—on board balloons, rockets, and, most importantly, satellites. Only as recently as the 1950s has this new research become possible.

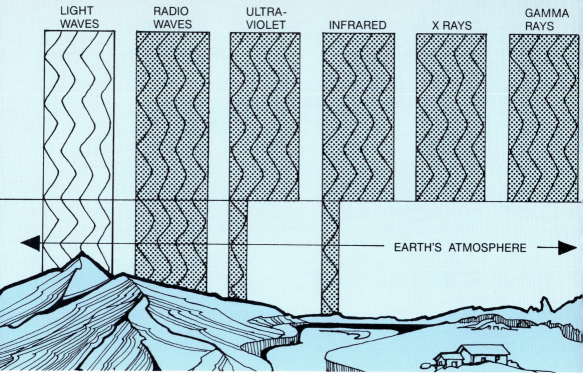

| LIGHT WAVES | RADIO WAVES | ULTRA-VIOLET | INFRARED | X RAYS | GAMMA RAYS |

EARTH'S ATMOSPHERE →

The earth's atmosphere has different effects on electromagnetic radiation of various wavelengths. Light waves and radio waves pass through, while X rays and gamma rays and most ultraviolet and infrared are blocked out.

Today, using various instruments on the ground and in space, scientists can study signals across the entire **electromagnetic spectrum.** From these studies they have learned about such strange objects as pulsars, **quasars, X-ray stars,** and **radio galaxies.** They have learned that there are places hotter and denser than we imagined possible. And, they have discovered that the universe as a whole is not calm and peaceful, but violent and ever-changing.

A TELESCOPE AT CERRO TOLOLO INTER-AMERICAN OBSERVATORY

2 More Light!

For hundreds of years, scientists had only their eyes with which to observe the universe. Then, just after the start of the seventeenth century, the telescope was invented. Soon, the Italian scientist Galileo Galilei used it to discover the four main moons of Jupiter, as well as the craters and mountains of our moon. Galileo also studied the **Milky Way.** He discovered that it was made up of stars "so numerous as to be almost beyond belief."

At first, telescopes were small. They used only lenses —curved pieces of clear glass—to form images of objects in space.

Later, telescopes became bigger. Because of their larger lenses, they could gather more light and show objects that were fainter. At the same time, some telescopes were built using a curved mirror in place of lenses. Today, all of our big, new telescopes are of this type.

Telescopes pick up great quantities of light from small parts of the sky. In addition, they focus this light, then magnify the image, so that we can look in detail at objects that are very dim or small.

By themselves, telescopes are powerful tools for studying the universe in ordinary light. But teamed with other instruments they are able to do even more.

Photography and Spectroscopy

Although telescopes are very good at gathering light, they are, by themselves, no better than the human eye at storing it. For this reason, in the late nineteenth century, astronomers began using telescopes as giant cameras. Images of still fainter objects—distant galaxies, for example—could then be made by exposing photographic plates for a long enough time.

Also in the nineteenth century came the **spectroscope,** an instrument that allowed very clear, precise spectrums to be obtained. Teamed with the telescope, the spectroscope allowed scientists to study the light coming from a single star, from glowing clouds of gas, and from entire galaxies. At last, the hidden signals in starlight were revealed.

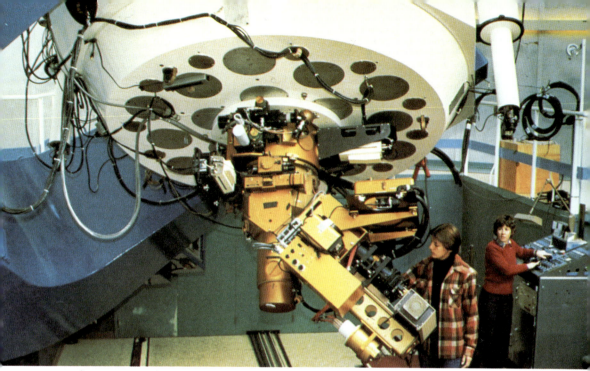

Astronomers check out the equipment at the 84-inch (2.1 meter) telescope at Kitt Peak National Observatory. The movements of most of the giant telescopes are controlled by computers. Some may be operated by a scientist thousands of miles away.

Telescopes, Present and Future

Today, the largest telescopes in the world have mirrors measuring more than 150 inches (380 centimeters) in diameter. They have equipment capable of sensing individual **photons,** or particles of light. The movements of these giant instruments are controlled by computers. They may even be operated from thousands of miles away by a scientist who watches the results on a television screen!

Often, telescopes are found in strange and remote places. They are atop extinct volcanoes in Hawaii and the Canary Islands, at the peak of snow-capped mountains in Europe, and in the remote outback of Australia. All these places were chosen because their near-perfect skies are well suited for viewing the heavens.

This aerial view of Cerro Tololo Inter-American Observatory in Chile shows its location on top of Cerro Tololo Mountain. The observatory has several telescopes, the largest of which has a mirror 157 inches (400 centimeters) in diameter. Astronomers at the observatory study high energy stars that give off strong radio waves and X rays, and the central parts of the Milky Way Galaxy.

Often, too, the telescopes themselves are strange to look at. In their search for greater and greater light-gathering power, scientists are designing instruments with huge single mirrors—40 feet (12 meters) or more across. They are designing others with several mirrors that can work together. One of these, the Multiple Mirror Telescope at Kitt Peak, Arizona, has already been built.

In the future we can expect larger telescopes on the ground with still better equipment for making use of the light they collect. Most exciting of all, though, we can look forward to giant telescopes in space!

Nowhere is there a better place from which to see the universe than above the earth's atmosphere. Here there are no clouds, no blurring caused by shifting layers of air —only a sky that is always perfectly dark and clear.

24

This is an artist's view of the Space Telescope as it is launched from a space shuttle orbiter. Future astronomy satellites such as this one will be able to study objects in space far more distant than those seen by instruments of earth.

With the launch of the Space Telescope by the space shuttle in late 1986, astronomy will take a great leap forward. Suddenly, there will be an instrument that can see objects 100 times fainter than any seen before. We can only imagine what it may discover.

The Universe of Light

Like all forms of electromagnetic radiation, light is given off when tiny particles, called **electrons,** lose energy. In addition, scientists describe electromagnetic radiation as either **thermal** or **nonthermal.**

Thermal radiation is given off by all objects simply because of their heat content. In general, the hotter an object, the shorter will be the wavelength of most of the energy that comes from it. Many normal stars, for in-

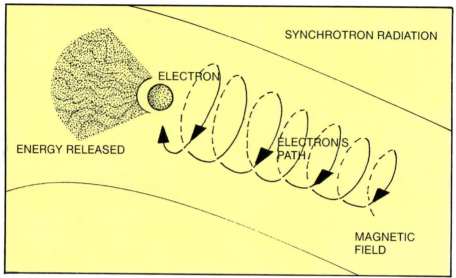

In this drawing, a charged particle—an electron—loses energy as it moves along at high speed through a magnetic field. In space, this type of nonthermal radiation is called synchrotron radiation.

stance, including the sun, give off most of their energy as visible light. Cooler objects radiate mainly infrared or radio waves. Hotter objects give off mostly ultraviolet or X rays.

Nonthermal radiation, however, has nothing to do with an object's temperature. In space, it most often comes from electrons losing energy while moving at high speed in a magnetic field. In this case, it is called **synchrotron radiation.**

At visible wavelengths, most of the radiation we see is thermal. It usually comes from peaceful regions of the universe: from normal stars, from glowing clouds of gas between the stars, and from entire cities of stars, or galaxies.

Yet, visible light spans only a tiny part of the full

Thermal radiation is given off by objects in space such as this spiral galaxy in the constellation of Canes Venatici. At visible wavelengths, most of the radiation we see is thermal, which is given off by objects simply because of their heat content.

electromagnetic spectrum. What strange objects, then, might we see at other, invisible wavelengths?

A LARGE RADIO TELESCOPE AT THE NATIONAL RADIO ASTRONOMY OBSERVATORY IN GREEN BANK, WEST VIRGINIA

3 **The Radio Revolution**

We can think of the earth's atmosphere as having two big windows, each giving a quite different view of space. Through the first, the "optical window," we see the universe in ordinary light. Through the second, the "radio window," we see an entirely new and exciting picture of space in radio waves.

The radio window lets through a much broader range of wavelengths than the optical window. Yet, the science of radio astronomy did not really begin until 1930. In that year Karl Jansky, an American telegraph engineer, by chance picked up radio waves coming from the sky. The source of these strange signals, it turned out, was hot gas near the middle of the **Galaxy.** Suddenly, the importance of the radio window to astronomy became clear.

Through Giant Eyes

In the 1950s, scientists built the first of the very large telescopes needed to study the universe through the radio window. Since radio waves range from about a thousand to many millions of times the length of light waves, the collecting "dish" of a radio telescope must be much bigger than the mirror of an ordinary telescope. Only in this way can the radio telescope "see" objects in space in reasonably fine detail.

Today, the largest radio telescope that can be fully steered, at Efflesburg in West Germany, is 328 feet (100 meters) in diameter. But this giant is small compared with the dish of the Arecibo Ionospheric Observatory in Puerto Rico. Making use of a natural basin in the ground, the collector at Arecibo measures about 1,000 feet (more than 300 meters) from rim to rim! At the same time, even this great instrument cannot see as clearly as the largest optical telescopes. Arecibo's giant "eye" has an ability to see detail—a **resolving power**—about 100 times less than that of an instrument working at visible wavelengths.

To boost resolving power in the radio region, astronomers use **interferometers.** These are groups of widely

The Very Large Array in New Mexico is an interferometer—widely spaced groups of radio telescopes whose observations can be combined to give a detailed view of the radio sky. Results from these telescopes are fed into computers to create a "radio map" of a small region of the sky.

spaced radio telescopes whose separate observations can be combined to give a much finer view of the radio sky. An example is the Very Large Array (VLA) in New Mexico: a huge Y-shaped arrangement of 27 telescopes, each with an 82-foot (25-meter) collecting dish. Results from the telescopes are fed into computers to create a "radio map" of a small region of the sky. Such a map may show detail as fine as that in the best optical photographs.

Sometimes interferometers are set up so that the individual instruments lie hundreds or thousands of miles apart. Radio telescopes on almost opposite sides of the earth have been teamed together. Through this approach, called Very Long Baseline Interferometry, the most detailed of all studies of radio sources in space can be carried out.

Dwellers of the Radio Sky

What a surprise was in store for radio astronomers in the 1950s and 1960s! Until then, the universe had seemed to contain only the normal stars and clouds of gas and dust that make up normal galaxies. Radio astronomy, though, soon showed that there were other, far stranger objects in space that had never before been suspected.

First to be found were the radio galaxies. Appearing as tiny, intensely bright specks in the radio sky, these puzzled astronomers for several years. Then, as the resolving power of radio telescopes improved, optical telescopes could be used to search for visible objects at the position of each radio source. In this way, the nature of radio galaxies became clear.

This unusual type of galaxy in the constellation of Centaurus is a radio source. Such a radio galaxy may give off a million times more radio waves than a normal galaxy like our own.

Through an ordinary telescope, most radio galaxies look like **giant elliptical galaxies**—the largest and heaviest type of star city known. Radio galaxies may give off a **million** times more radio waves than a normal galaxy like our own. Stranger still, this great radio energy does not come from the visible part of the galaxy where there are the most stars. Instead, it usually comes from two huge radio "lobes" located on either side. Astronomers know that these lobes must be filled with very hot gas, called **plasma,** and that their powerful radio emissions are created by electrons in a magnetic field. But they do not know how the electrons and the rest of the plasma are shot out in such vast quantity, and at such high speed, from the main part of the galaxy.

One theory, now gaining strength, is that at the very

33

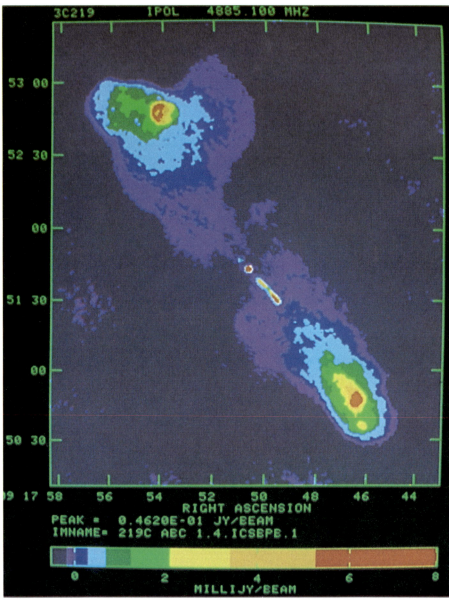

This false-color image shows the extended lobes, hot spots, and a short jet extending from the core of the radio galaxy 3C-219. The hot spots are believed to be the emissions from high-pressure regions formed at the end of jets that transport energy from the core of the galaxy to the lobes.

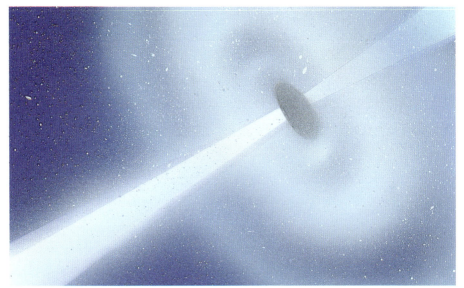

In this picture, an artist imagines twin jets of hot matter shooting out in opposite directions from a black hole in the center of an active galaxy.

center of each radio galaxy there is an enormous **black hole**—a region of space where gravity is so strong that nothing, not even light, can escape. Such a black hole would weigh several hundred million times as much as the sun. It would then be possible for hot matter, approaching close to the black hole, to be squirted out in opposite directions along the black hole's spin **axis.**

In 1963 astronomers found some objects even more startling than radio galaxies. These were quickly given the name quasars, from "quasi-stellar objects," since through optical telescopes they look like ordinary blue stars. The quasars, however, are still more distant and powerful than radio galaxies. Over all wavelengths (not just radio waves), a quasar may shine as brightly as a hundred normal galaxies. What is more, all of this energy

This unusual irregular galaxy in the constellation of Ursa Major is an example of an active galaxy. Scientists believe that such active galaxies may once have been quasars, and that quasars are the bright cores of very young galaxies.

seems to pour out from a tiny region at the heart of the quasar no bigger than our solar system! Again, astronomers have turned to the idea of a "supermassive" black hole to explain this unusual behavior.

Gradually, the link between quasars and other powerful sources such as radio galaxies is becoming clearer. Quasars are, almost certainly, the bright cores of very young galaxies—galaxies being seen now as they were when the universe itself was new. As the quasars aged, less matter may have fueled their central black holes. In time, they may have turned into less violent forms, including radio galaxies and other types of **active galaxies.**

In other areas radio astronomers have made equally dramatic breakthroughs. One discovery was radio waves

In this picture, a pulsar—a flashing neutron star—turns on its axis. As it turns, it sends out a steady stream of short, sharp radio "beeps" or pulses. Like a lighthouse, the pulsar sends out radiation along two narrow beams that flash on and then off.

with a certain special wavelength. Called **21-centimeter radiation**, these waves are given off by clouds of cold hydrogen. Their discovery made possible the first detailed mapping of the spiral arms in the Galaxy. Then, in 1967, came the unexpected finding of pulsars.

A pulsar sends out a steady stream of short, sharp radio "beeps" or pulses. These pulses are extremely regular, like the ticks of a very accurate clock. The time between them, depending on the individual pulsar, may be from less than a hundreth of a second to several seconds.

Because a pulsar is small, spins rapidly, and has a powerful magnetic field, it sends out radiation like a lighthouse along two narrow beams. The pulses we see are caused by these beams flashing across our line of

The Crab nebula is a gassy supernova remnant—the remains of a supernova seen from earth in 1054.

sight. As the pulsar turns, they flash on and then off.

Usually a pulsar forms when a big, heavy star explodes, at the end of its life, as a **supernova**. What is left of the big star's core may shrink to become a **neutron star**—a solid ball of **neutrons** less than 20 miles (32 kilometers) in diameter. A pulsar is just a flashing neutron star.

The most famous of all pulsars lies at the center of the Crab nebula—the remains of a supernova seen from earth in 1054. It is one of only two pulsars known to flash in ordinary light.

Astronomers are interested, too, in the gassy **supernova remnants**. These scattered star wrecks still shine brilliantly, especially at radio and X-ray wavelengths, by synchrotron radiation.

Scientists used the Holmdel Horn Antenna in New Jersey to discover the cosmic microwave background.

The Noise of Creation

About 15 **billion** years ago, scientists believe, the universe began in an incredibly powerful explosion called the **Big Bang**. At first, space was filled with a very hot "soup" of particles and energy. Today, we can see the glow of radiation from this early soup. It appears as a steady hiss of radio waves at short wavelengths—the so-called **microwave background.**

With wavelengths between about 39 inches (1 meter) and 39 thousandths of an inch (1 millimeter), **microwaves** make up the part of the electromagnetic spectrum between longer wavelength radio waves and infrared rays. Astronomers have found that many **molecules** in space—including those of substances such as water, ammonia, and alcohol—give off microwaves.

THE FIRST X-RAY SATELLITE, *UHURU*

4 The View From Space

Today, astronomy is growing faster than ever before. Skilled scientists and engineers have learned how to overcome the barrier of the earth's atmosphere. Now, as well as using optical and radio telescopes on the ground, they have sent instruments to the edge of space and into earth orbit. Space programs have led to the birth and development of new fields—X-ray, gamma ray, infrared, and ultraviolet astronomy. These new areas of knowledge have given us a much deeper understanding of the universe.

X-ray Astronomy

If we could look into space with X-ray eyes, what would we see? Instruments carried aboard **sounding rockets** and satellites have already given us some of the answers. X-ray eyes would show regions in the universe of great energy. In these mysterious places, the temperature is unimaginably high or particles race close to the speed of light.

X-ray astronomy took a giant step forward in December 1970 with the launch of the first X-ray satellite, *Uhuru.* Since then, other orbiting spacecraft, such as *Ariel V, Einstein* (the first High Energy Astrophysical

Observatory), and *Exosat,* have added greatly to our knowledge of the X-ray sky.

Satellites have found new types of objects in space; for example, the **X-ray binary.** Imagine two stars circling closely about each other. One of the pair is a normal star, large and bright. The other is a **collapsed star,** which may be either a neutron star, or, possibly, a black hole. Gas from the normal star is sucked by **gravity** into a fiery whirlpool—an **accretion disk**—centered on the collapsed companion. Here, the gas is heated to a temperature of several million degrees. At this great heat, it begins to give off X rays.

Because the amount of hot gas in the accretion disk varies from one moment to the next, an X-ray binary flickers rapidly. Such an object is called a **transient source** of

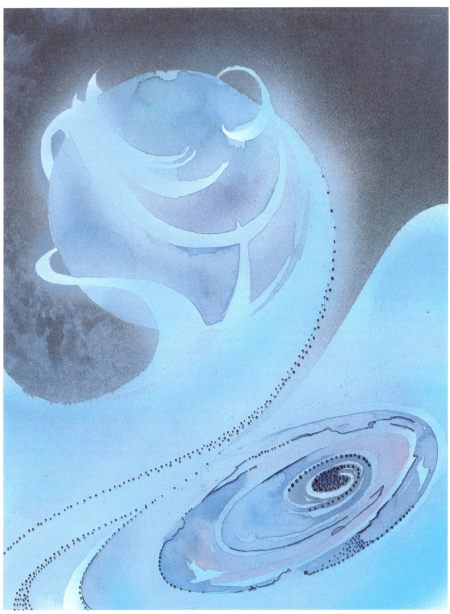

This is an artist's view of Cygnus X-1, an X-ray binary star that probably contains a black hole. The black hole is a collapsed star that can strip away huge amounts of matter from a normal star close by.

radiation—a source that is constantly changing.

The most well known X-ray binary is **Cygnus X-1.** Recent studies of its X rays and of its collapsed companion have produced an exciting discovery. Cygnus X-1 almost certainly contains a black hole!

Much larger black holes may lurk within the cores of quasars and active galaxies. In some cases these objects have also turned out to be transient X-ray sources. Their flickering X rays support the idea of a small, central "powerhouse" in young, active galaxies. This power source is probably a black hole the size of our solar system surrounded by a swirling accretion disk.

Some X-ray sources flare up so suddenly that they have been named **X-ray bursters.** They are probably a special type of X-ray binary in which a gradual buildup,

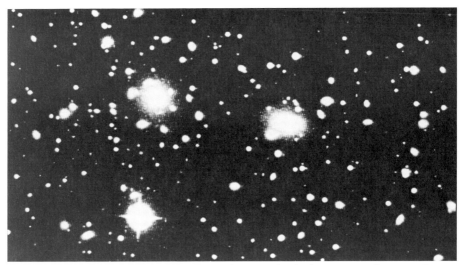

X-ray astronomers have looked at large clusters of galaxies at great distances from earth. In clusters of galaxies such as this one in the constellation of Coma Berenices, they have found X-ray-emitting gas. Such gas exists in the vast spaces between galaxies.

and then a sudden explosion, of matter occurs on the surface of a neutron star.

Also at great distances, X-ray astronomers have looked at large **clusters of galaxies.** Within them, they have found superhot, X-ray-emitting gas—gas that exists not in the galaxies themselves, but in the immense spaces between them.

Gamma Ray Astronomy

At even shorter wavelengths and higher energies than X rays are gamma rays. Could these reveal still more about the most powerful objects in the universe? To find the answer, astronomers must solve a number of big problems.

First, gamma rays usually behave, not like waves, but

like "bullets" of energy. Special instruments called **particle detectors** are needed to study them. Particle detectors, though, have a very low resolving power. This makes it difficult to pinpoint the position of gamma ray sources in the sky.

Gamma ray sources are also scarce. Only about 30 are known at present. Also, the number of gamma rays reaching the earth is tiny compared with the number of light waves or radio waves.

Much work in gamma ray astronomy has been done with just two satellites, *SAS-2* and *COS-B*, along with high-flying balloons. Some of the gamma ray sources found are already well known at other wavelengths. They include the Crab pulsar, an X-ray binary, a handful of active galaxies, and a quasar. Other gamma ray sources are

The Tarantula nebula in the constellation of Dorado contains a central cluster of hot, blue supergiant stars. This central cluster gives off most of its energy as ultraviolet radiation.

new. **Gamma ray bursters,** for example, are similar to X-ray bursters, but of even higher energy. Regions where **cosmic rays** bump into clouds of gas in the Galaxy have also been seen for the first time through gamma ray observation.

Gamma ray astronomy lets us peer deep inside the highest energy objects in space. Through this exciting new field, we may explore a young pulsar, the nucleus of a quasar, or the places where fast-moving particles smash into others and shatter.

Ultraviolet Astronomy

Balloons, rockets, and satellites are also used to see the universe at ultraviolet wavelengths. Spacecraft such as the *Orbiting Astronomical Observatories* and the

International Ultraviolet Explorer have been especially important.

Viewed in ultraviolet, the brightest objects in the sky are hot stars. If the surface temperature of a star is greater than about 18,000°F (10,000°C), then it will give off most of its energy as ultraviolet radiation.

Many ultraviolet stars are young, including the giant **OB types.** These are the brightest and hottest stars in the Galaxy. Others are older; for example, the hot **subdwarfs,** the central stars of **planetary nebulas,** and the so-called **Wolf-Rayet stars.**

Infrared Astronomy

Some infrared astronomy can be done at ground level. In this case, though, the observatory must be in a

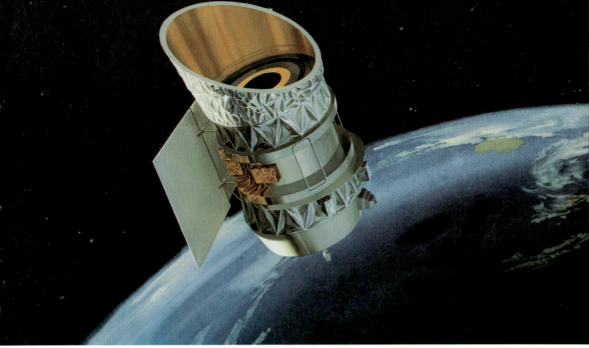

In this view, an artist shows the Infrared Astronomy Satellite in earth orbit. Launched on January 25, 1983, this orbiting telescope conducted a ten-month, all-sky survey of the infrared sky. It pinpointed stars, galaxies, and clouds of dust and gas that give off infrared energy.

place that is unusually high and dry, since infrared rays are absorbed by **water vapor** in the atmosphere. One of the best observing sites on earth is the peak of Mauna Kea in Hawaii. Perched high on this volcanic mountain are the two largest infrared telescopes in the world.

Infrared radiation covers a broad wavelength band from around 30 millionths of an inch (800 nanometers) to 4 hundredths of an inch (1 millimeter). Only a tiny part, though, can be viewed from the ground. To see across the full infrared spectrum requires instruments in space—a need which led, in 1983, to the launch of the *Infrared Astronomy Satellite.*

What then, do we see when we look into the infrared universe? Just as ultraviolet comes from very hot objects, so infrared comes from warm objects. The most common

The Horsehead nebula in the constellation of Orion is an example of an interstellar gas and dust cloud. Such warm regions in space are often the birthplaces of new stars.

sources of infrared in space are clouds of gas and dust with temperatures as high as a few hundred degrees. Often, these are the birthplaces of new stars. Infrared astronomy, in fact, can teach us much about how stars are made.

In other galaxies, infrared studies have shown "starbursts" taking place. In a starburst, many stars are formed together in one area of space. Such an event is often brought about when two galaxies approach very closely or run into each other.

As with radiation at most other wavelengths, infrared pours out in great quantities from the centers of quasars and active galaxies. Recently, astronomers have discovered something unusual in the core of our own galaxy. At infrared (and radio) wavelengths, a strong source

Nearly the entire sky, as seen in infrared wavelengths, is shown in this image put together from six months of Infrared Astronomy Satellite data. The bright yellow and red band running across the picture is the plane, or disk, of the Milky Way Galaxy. The center of the Galaxy is located at the center of the picture.

has been found that, so far, is unexplained. It may be dust heated by neighboring stars, or it may be radiation coming from around a large black hole. Perhaps within a few years we shall know for sure!

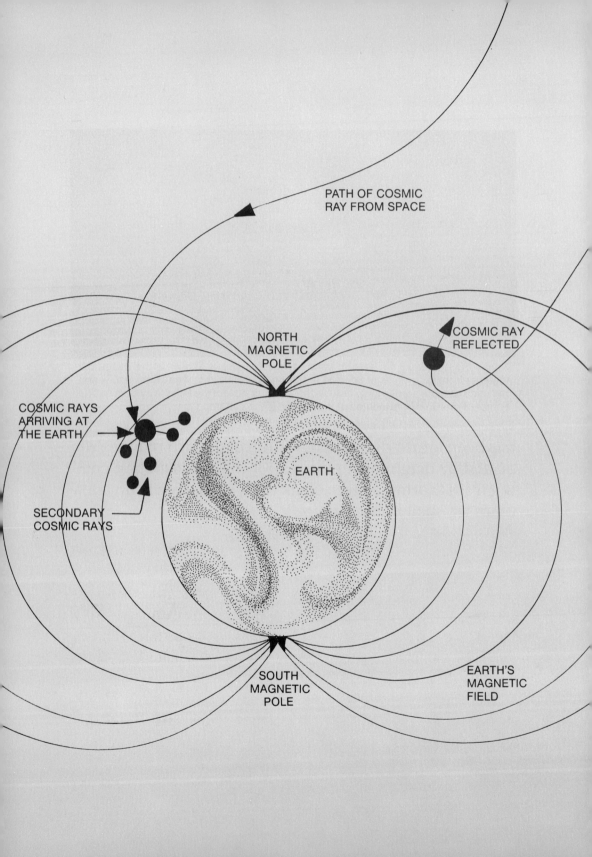

PATH OF COSMIC
RAY FROM SPACE

COSMIC RAY
REFLECTED

NORTH
MAGNETIC
POLE

COSMIC RAYS
ARRIVING AT
THE EARTH

EARTH

SECONDARY
COSMIC RAYS

SOUTH
MAGNETIC
POLE

EARTH'S
MAGNETIC
FIELD

5 Into the Unknown

Studying electromagnetic radiation, in all its forms, is the best way scientists have of learning about the universe. But it is not the only way.

Near to home, **meteorites** that strike the earth provide clues to the early history of the solar system. Recently, too, astronauts and robot spacecraft have begun to explore nearby worlds.

From beyond the sun's kingdom, though, come messages not carried by electromagnetic waves. This strange, new information may reach us as cosmic rays, **neutrinos,** or **gravity waves.**

The Mystery of Cosmic Rays

Arriving at the earth from all directions in space are fast moving particles called cosmic rays. Most cosmic rays are high-speed pieces of **hydrogen** (**protons**) or pieces of heavier substances such as **helium.** As they move along, their paths are bent by the magnetic field of the Galaxy. This bending makes it very difficult to find out where cosmic rays come from.

When a cosmic ray crashes into the earth's atmosphere, it sends a shower of new particles, called "secondary" cosmic rays, raining down to the surface. Scientists, then, can study these secondary cosmic rays on the ground. They can also look for "primary" cosmic rays in space.

Scientists have discovered that cosmic rays have a very wide range of energies. Those of lower energy may be made within the Galaxy and those of higher energy outside it. Likely sources of cosmic rays include pulsars, supernovas, X-ray binaries, active galaxies, and quasars.

Neutrinos and Gravity Waves

Every second, trillions of strange, ghostlike particles pass through our bodies, through the earth, and then speed out into space again. These are neutrinos, particles that travel at the speed of light and are hardly ever stopped by ordinary matter.

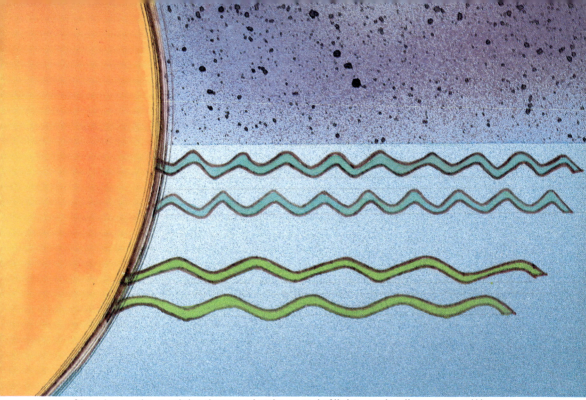

Neutrinos—tiny particles that travel at the speed of light—and radio waves and X rays are given off by the sun and reach or pass through the earth.

One important nearby source of neutrinos is the sun. Neutrinos, flowing out from the sun's core, may teach us much about how our neighborhood star makes its energy. To study the sun's neutrinos, a special detector—an enormous tank of cleaning fluid—has been set up deep inside a South Dakota gold mine. Only neutrinos, of all the particles from space, can pass through the thick overlying rock and reach the detector.

Neutrino astronomy is still a new area of study. Scientists are just starting to build detectors to search for neutrinos coming from distant parts of space—from such objects as supernovas and quasars—and from cosmic rays that crash into the earth's atmosphere. At the same time, they are searching for an even stranger kind of space signal.

To study neutrinos, the Brookhaven solar neutrino experiment was set up 4,900 feet underground in the Homestake Gold Mine in Lead, South Dakota. The neutrinos are captured in a huge tank filled with cleaning fluid. This experiment is located deep below the earth's surface so that cosmic rays and all other particles except the neutrinos will be shielded out by the earth.

When any object in the universe with **mass** moves, it should make "ripples" in the surrounding space. These unusual ripples are called gravity waves. Like electromagnetic waves, they should be able to tell us much about the places from which they have come. Before they can be studied, though, there is a problem to be solved.

Gravity waves, even from the strongest sources, are incredibly weak. Detecting them has so far proved impossible, and a successful gravity wave detector may still lie a number of years in the future.

In the meantime, scientists have discovered a handful of objects in space that serve to test their ideas about how gravity waves work. These objects are called binary pulsars. Made of two heavy stars circling closely about each other, one of which is a pulsar, they are an ideal

In this picture, an artist imagines a binary pulsar—a twin star made of a pulsar and an ordinary heavy star. Scientists are using these objects to study gravity waves.

natural laboratory for studying gravity.

If the stars are losing energy, as expected, by sending out gravity waves, they should gradually be taking longer to complete each orbit. In fact, this is just what scientists have found. The clocklike ticks of the pulsar allow the slowing down to be timed very accurately.

Looking to the Future

The new astronomy of recent years has taught us that we live in an extraordinary universe. Through "eyes" that see invisible waves and strange particles, astronomers have discovered objects of very high energy and density—objects that change in brightness from one instant to the next. What fantastic things will they find in the years to come?

Sunset dimly shows the shape of domes at Kitt Peak National Observatory. In the future, supertelescopes on earth and in space may lead to many new astronomical discoveries.

Already, scientists are building telescopes far more powerful than any in use today. Soon there will be supertelescopes, both on the ground and in space, for studying X rays and gamma rays, infrared and ultraviolet, radio waves and ordinary light. In addition, there will be new instruments that can read more clearly the messages carried by cosmic rays, neutrinos, and gravity waves. The future of astronomy and space exploration is exciting indeed—a future in which you can take part.

Appendix A:
Discover For Yourself

1. *Observatories*

There is no better way to learn, and sense the excitement, of modern astronomy than by visiting an observatory. In the United States, and in other parts of the world, a number of large observatories offer services to the public. Telescopes may be on display, and there is usually an exhibition set up to explain the observatory's work.

Among observatories that welcome visitors are:

Hale Observatories, Pasadena, California

Kitt Peak National Observatory, Tucson, Arizona (including the KPNO Visitor Center)

Mauna Kea observatories, Hawaii (a tour of several observatories on this site can be arranged through Mauna Kea Observatory Support Services, 177 Makaala Street, Hilo, Hawaii 96720; telephone (808) 935-3371)

National Radio Astronomy Observatory, Green Bank, West Virginia

Yerkes Observatory, Williams Bay, Wisconsin

Very Large Array National Observatory, Socorro, New Mexico

2. *Other Places of Interest*

A visit to a planetarium is also a good way to keep up with new developments in space. Find out where your nearest planetarium is located, and obtain a copy of its current program.

In some parts of the country, there are special space exhibitions and museums. Here are just a few:

Lawrence Hall of Science, University of California, Berkeley ("hands on" astronomy exhibits and a planetarium)

Johnson Space Center, Houston, Texas (spacecraft exhibits)

National Air and Space Museum, Washington, D.C. (aerospace exhibits and a planetarium)

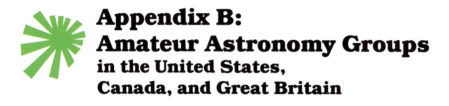

Appendix B:
Amateur Astronomy Groups
in the United States,
Canada, and Great Britain

For information or resource materials about the subjects covered in this book, contact your local astronomy group, science museum, or planetarium. You may also write to one of the national amateur astronomy groups listed below.

United States
The Astronomical League
Donald Archer,
   Executive Secretary
P.O. Box 12821
Tucson, Arizona 85732

American Association of
   Variable Star Astronomers
187 Concord Avenue
Cambridge, Massachusetts 02138

Canada
The Royal Astronomical Society of Canada
La Société Royale d'Astronomie du Canada
Rosemary Freeman, Executive Secretary
136 Dupont Street
Toronto, Ontario M5R 1V2

Great Britain
Junior Astronomical Society
58 Vaughan Gardens
Ilford
Essex IG1 3PD England

British Astronomical Assoc.
Burlington House
Piccadilly
London W1V 0NL England

Glossary

accretion disk—a whirlpool of hot matter swirling around a collapsed star, such as a black hole

active galaxy—a type of galaxy that gives off much more energy than an ordinary galaxy like our own

axis—the imaginary line about which a spinning object seems to turn

Big Bang—the great explosion in which, scientists believe, our universe began

billion—a thousand million. Written as 1,000,000,000

black hole—a region of space where the pull of gravity is so strong that nothing, not even light, can escape

cluster of galaxies—a group of galaxies, with from a few dozen to more than a thousand members, that is held together by gravity

collapsed star—a star that, at the end of its life, becomes very small and dense as a result of being squeezed by gravity

cosmic rays—particles of very high energy that reach the earth from all directions in space. They are mostly protons, together with heavier particles, electrons, and gamma rays

Crab nebula—the bright, gassy remains of a star that was seen to explode as a supernova in 1054. It is a source of all types of electromagnetic radiation and one of the most interesting space objects known

Cygnus X-1—a strong source of X rays. It is thought to be a binary system in which one of the

stars is a black hole

density—a measure of how concentrated matter is. It tells how much mass there is within a given volume

diameter—the length of a straight line that runs from one side of an object to the other, passing through its center

electromagnetic spectrum—the complete range of wavelengths over which electromagnetic waves exist

electromagnetic waves—waves of vibrating electric and magnetic fields. The various types of electromagnetic waves differ only in their wavelengths, and include: radio waves, infrared, visible light, ultraviolet, X rays, and gamma rays

electron—a tiny particle, even smaller and lighter than a proton or a neutron

galaxy—a star city, held together by its own gravity

Galaxy, the—the star city—a large spiral galaxy—in which we live

gamma ray burster—an object that sends out a sudden powerful burst of gamma rays. A gamma ray burster may be similar to an X-ray burster but is of even greater energy

gamma rays—electromagnetic waves with the shortest wavelengths (and highest energies) of all

giant elliptical galaxy—the largest type of star city known. Giant elliptical galaxies are ball-shaped, may contain more than a trillion stars, and are sometimes a

powerful source of radio waves

gravity—the force by which all objects with mass pull on all other objects with mass

gravity waves—ripples in space believed to be caused by any massive moving object. Gravity waves travel with the speed of light, are extremely weak, and, so far, remain undetected

helium—a gas, and the second lightest substance in the universe

hydrogen—a gas. The lightest substance of all. It makes up three-quarters of the matter in most stars and is the main fuel for nuclear fusion in their cores

infrared rays—the kind of electromagnetic waves that carry heat. They are given off by all warm objects

interferometer—an instrument, made of two or more linked radio telescopes, that allows radio sources to be studied in fine detail

magnetic field—the area of space near a magnetic body such as the earth

mass—the amount of matter in a body as measured by its ability to stay at rest or move in the same direction

matter—anything that can have weight and takes up space

meteorite—a rock from space that strikes the earth's surface

microwave background—the steady "hiss" of microwaves coming from all parts of the sky. Scientists believe that the microwave background is radiation reaching us from a time shortly after the Big Bang

microwaves—radio waves of short wavelength

Milky Way—the thin, ragged band of misty light that stretches across our sky from north to south. Made of distant stars, it is part of the disk of our galaxy seen edge-on

million—a thousand thousand. Written as 1,000,000

millionth—one divided by a million

molecules—small groups of particles of which most substances are made. Clouds of molecules account for roughly half of all the gas among the stars

neutrino—a tiny, ghostlike particle that travels at the speed of light. Countless numbers of neutrinos reach us from the sun and other objects in space

neutron—a small particle, similar to the proton

neutron star—a type of collapsed star made of neutrons packed closely together. A neutron star measures only about 20 miles (32 kilometers) in diameter but weighs at least 1½ times as much as the sun

nonthermal radiation—electromagnetic radiation that does not depend on the temperature of its source

OB-type star—an O-type or B-type star—the brightest, hottest types of stars in the Galaxy

particle detector—an instrument used to detect tiny particles, such as protons, or high-energy waves, such as gamma rays

photon—electromagnetic radiation can behave either as waves or as particles. A particle of electromagnetic radiation is called a photon

planetary nebula—a glowing ring, or shell, of gas thrown off by stars like the sun as they grow old

plasma—gas so hot that it has broken up into small pieces including electrons, and, possibly, protons and neutrons

prism—a triangular block of glass that can split white light up into the colors of the rainbow

proton—a small particle that, together with the neutron and electron, is one of the building blocks of ordinary matter

pulsar—a newly formed neutron star that, as it spins, sends out radiation along two narrow beams

quasar—the brightest, most distant type of object known. Quasars are probably the cores of young active galaxies

radio galaxy—a galaxy that gives off most of its energy as radio waves

radio telescope—an instrument used to pick up radio waves coming from space

radio waves—electromagnetic waves with the longest wavelengths (and lowest energies) of all

resolving power—the ability of a telescope (or

other instrument) to make out detail in a source of radiation

sounding rocket—a small rocket that can carry instruments, for a short time, high into the earth's atmosphere

spectroscope—an instrument used to study a spectrum in fine detail

spectrum—the result of spreading out a beam of radiation, such as light, into its various wavelengths

subdwarf—an unusually small star (though not a collapsed star)

supernova—the explosion of a heavy star. It results in a supernova remnant and, possibly, either a neutron star or a black hole

supernova remnant—the glowing cloud of gas resulting from a supernova

synchrotron radiation—electromagnetic radiation (of any wavelength) that comes from electrons moving at high speed in a magnetic field

thermal radiation—electromagnetic radiation given off because of an object's heat content. The wavelength of thermal radiation depends on the object's temperature

transient source—a source of radiation whose output changes very rapidly

trillion—a million million. Written as 1,000,000,000,000

21-centimeter radiation—radio waves of an exact wavelength given off by cold clouds of hydrogen in space

ultraviolet rays—electromagnetic radiation with wavelengths between that of visible light and X rays

water vapor—the gas formed from water. On earth, it is the substance of which most clouds are made

wavelength—the distance between two successive crests in a train of waves

Wolf-Rayet star—an unusual type of hot, bright star that loses matter quickly into the surrounding space

X-ray binary—two stars—one of which is a normal star, the other a collapsed star—that are circling closely about each other and producing X rays. The X rays come from a whirl-pool of hot gas centered on the collapsed star

X-ray burster—an X-ray source that sends out short, sharp bursts of X rays. An X-ray burster probably consists of a neutron star circling closely around a normal star

X rays—electromagnetic radiation with wavelengths between those of ultraviolet rays and gamma rays

X-ray star—see *X-ray binary*

Suggested Reading

Asimov, Isaac. *Eyes on the Universe.* Boston: Houghton Mifflin, 1975.
A fascinating, detailed history of the telescope. Also included are chapters on radio astronomy and the recent use of satellites and balloons for work at short wavelengths. (Advanced)

Burnham, Robert. "IRAS and the Infrared Universe." *Astronomy,* March 1984, pp. 6-22.
Our first clear view of the universe in infrared came with the launching of IRAS, the Infrared Astronomy Satellite, in 1983. This well-illustrated article explains some of the amazing discoveries made by this spacecraft. (Advanced)

Henbest, Nigel. "Active Galaxies: Feeding the Dragon." From the *1985 Yearbook of Astronomy,* pp. 169-180. New York: W.W. Norton, 1984.
One of the strangest new findings of astronomy is that some very bright, unusual galaxies, together with quasars, may have giant black holes inside their cores. This article explains how such "supermassive" black holes, and their surrounding matter, could make active galaxies shine. (Advanced)

Knight, David C. *Eavesdropping on Space: The Quest of Radio Astronomy.* New York: William Morrow, 1975.
A good introduction to the principles of radio astronomy and to the exciting early discoveries made by this new science. (Intermediate)

Index

Milpitas Unified School District
Milpitas, California
Spangler School